Reproducing an Angle

Problem: *Reproduce a given angle.*

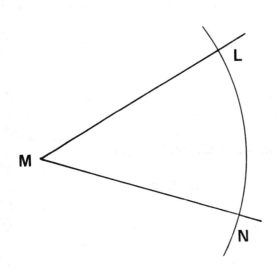
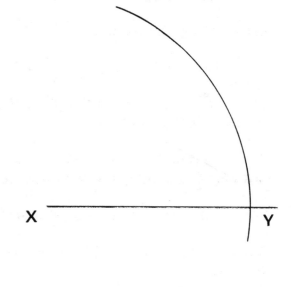

Solution:

1. Does the arc with center X have a radius congruent to that of the arc
   with center M? _ _ _ _ _

2. Draw an arc with center Y whose radius is congruent to segment $\overline{NL}$.
   (Make it intersect the arc through Y.)

3. Label as Z the point in which the arcs intersect.

4. Draw $\overrightarrow{XZ}$.

5. Is angle ZXY congruent to angle LMN? _ _ _ _ _ _
   How do you know? _
   _

Problem: *Reproduce a given angle.*

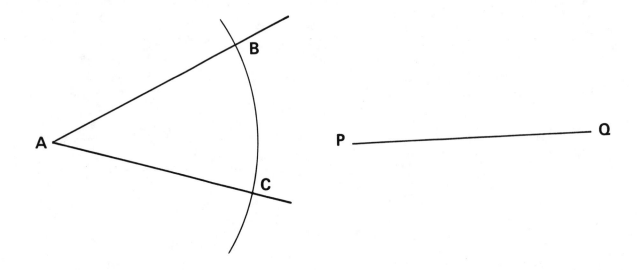

Solution:

1. Draw an arc with center P and radius congruent to segment $\overline{AC}$.

2. Label as R the point at which it intersects $\overrightarrow{PQ}$.

3. Draw a second arc with center R and radius congruent to $\overline{CB}$.

4. Label as S the point at which the second arc intersects the first arc.

5. Draw $\overrightarrow{PS}$.

6. Is angle BAC congruent to angle SPR? _ _ _ _ _

Key to
Geometry®

Angles

By Newton Hawley and Patrick Suppes
Revised by George Gearhart and Peter Rasmussen

Name _____ Class _____

KEY TO GEOMETRY
Book 6: Angles
TABLE OF CONTENTS

TO THE STUDENT:

These books will help you to discover for yourself many important relationships of geometry. Your tools will be the same as those used by the Greek mathematicians more than 2000 years ago. These tools are a **compass** and a **straightedge**. In addition, you will need a **sharpened pencil**. The lessons that follow will help you make drawings from which you may learn the most.

The answer books show **one** way the pages may be completed correctly. It is possible that your work is correct even though it is different. If your answer differs, re-read the instructions to make sure you followed them step by step. If you did, you are probably correct.

Cover art by Howard Coale.

Hypatia taught mathematics at the Greek university in Alexandria, Egypt, 700 years after Euclid. Hypatia was one of the university's most popular lecturers, and students came from Europe, Asia and Africa to study with her. She wrote On the Conics of Appolonius, a book about the curves that can be formed by slicing a cone: the circle, the ellipse, the parabola, and the hyperbola. By 400 A.D., the time of Hypatia, Roman armies had conquered Greece and controlled Alexandria. The Romans were very hostile towards Greeks and Greek ideas. During one outbreak of hostility, a mob brutally murdered Hypatia. Her death marked the end of the great age of Greek geometry.

On the cover of this booklet Hypatia studies models of the conic sections on a balcony overlooking Alexandria's harbor. (For more information about Hypatia and other women in mathematics, see Teri Perl's book, Math Equals, Addison Wesley, 1978.)

® Key to Fractions, Key to Decimals, Key to Percents, Key to Algebra, Key to Geometry, Key to Measurement, and Key to Metric Measurement are registered trademarks of Key Curriculum Press.
Published by Key Curriculum Press, 1150 65th Street, Emeryville, CA 94608
Printed in the United States of America   33 32 31 30   08 07 06 05 04   ISBN 0-913684-76-7

Problem: *Reproduce a given angle.*

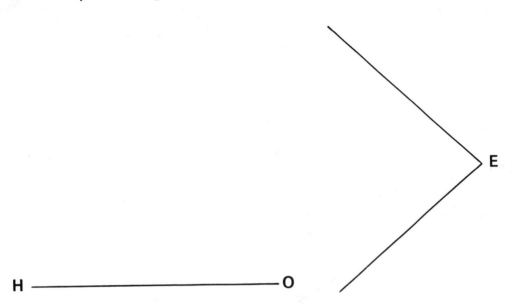

Solution:

1. Draw an arc with center E which intersects both sides of the angle.

2. Label as D and F the points of intersection.

3. Draw an arc of center O and radius congruent to segment $\overline{ED}$.

4. Label as B the point of intersection of this arc with ray $\overrightarrow{OH}$.

5. Draw an arc of center B and radius congruent to segment $\overline{DF}$.
   (Make it intersect the arc with center O.)

6. Label as G the point of intersection of the two arcs.

7. Draw the ray $\overrightarrow{OG}$.

Problem: *Construct an angle, with vertex W, which is congruent to angle POX.*

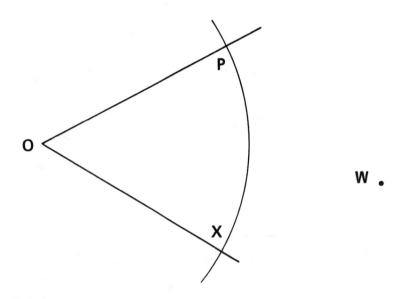

Solution:

1. Draw a line segment congruent to segment $\overline{OX}$ with W as one endpoint.

2. Label as T the other endpoint of the segment.

3. Draw an arc with center W and radius congruent to segment $\overline{OX}$. (Make it large. It should pass through T.)

4. Draw an arc with center T and radius congruent to segment $\overline{XP}$.

5. Label as U the point of intersection of these two arcs.

6. Draw $\overrightarrow{WU}$.

7. What is the vertex of angle TWU? _ _ _ _ _

Problem: *Construct an angle, with vertex Q, which is congruent to the given angle.*

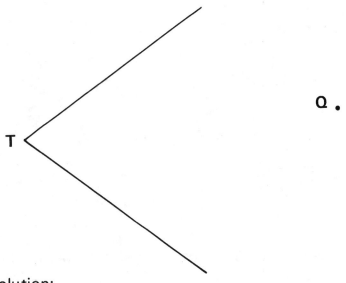

Solution:

1. Draw a ray with Q as endpoint.

2. Draw an arc with center T which intersects both sides of the given angle.

3. Label the points of intersection A and B.

4. Use the same radius to draw a large arc with center Q. (Make it intersect the ray.)

5. Label as R the point where the arc and ray intersect.

6. Draw an arc with center R and radius congruent to segment $\overline{AB}$. Make it intersect the arc of center Q.

7. Label the point where the arcs intersect as P.

8. Draw $\overrightarrow{QP}$.

9. Is angle PQR congruent to the given angle? _ _ _ _ _

1. Construct an angle, with vertex P, which is congruent to angle ABC.

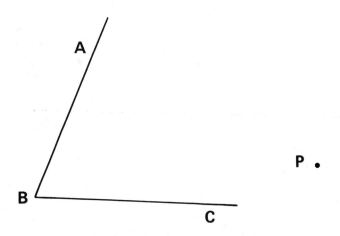

2. Construct an angle, with vertex M and ray $\overrightarrow{MN}$ as one side, which is congruent to angle XYZ.

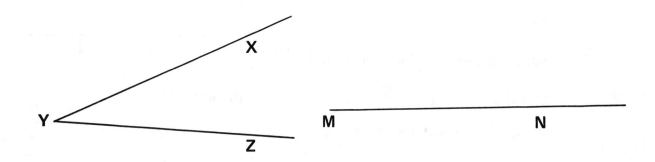

Review

1. On the given line, lay off a segment congruent to segment $\overline{PQ}$.

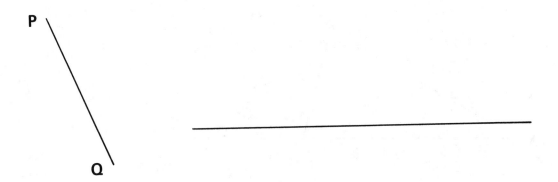

2. Construct an angle with vertex D and $\overrightarrow{DE}$ as one side which is congruent to angle ABC.

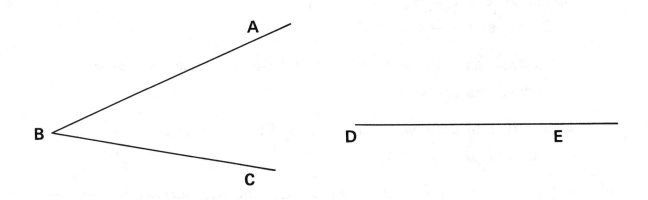

3. Bisect the given segment.

Reproducing a Triangle

Problem: *Reproduce a given triangle.*

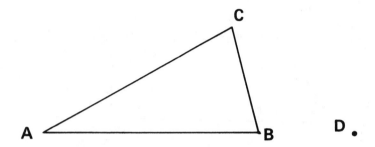

Solution:

1. Draw a line through D.

2. On the line, construct a segment with D as one endpoint which is congruent to segment $\overline{AB}$.
   Label the other endpoint E.

3. Construct an angle with vertex D and $\overrightarrow{DE}$ as one side which is congruent to angle BAC.

4. Construct an angle with vertex E and $\overrightarrow{ED}$ as one side which is congruent to angle CBA.

5. Label as F the point where the sides of the two angles intersect.

6. Is triangle DEF congruent to triangle ABC? _ _ _ _ _

1. Reproduce triangle XYZ.

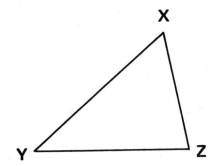

2. Reproduce triangle RST.

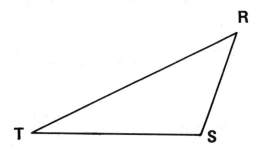

Problem: *Reproduce a given triangle.*

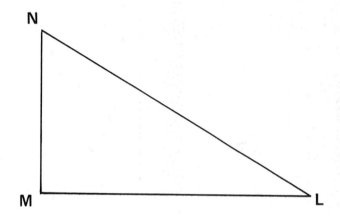

Solution:

1. Draw a segment congruent to segment $\overline{LM}$.
   Label its endpoints P and Q.

2. Construct an angle with vertex P and $\overrightarrow{PQ}$ as one side which is congruent to angle MLN.

3. Construct an angle with vertex Q and $\overrightarrow{QP}$ as one side which is congruent to angle LMN.

4. Label as R the intersection of the two angles.

5. Check: Is angle PRQ congruent to angle LNM? _ _ _ _ _

6. Check: Is side $\overline{PR}$ congruent to side $\overline{LN}$? _ _ _ _ _

   Check: Is side $\overline{QR}$ congruent to side $\overline{MN}$? _ _ _ _ _

Problem: *Reproduce a given triangle.*

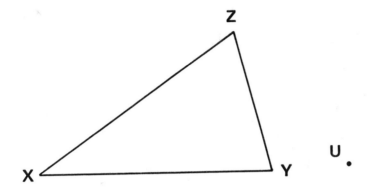

Solution:

1. Draw a segment congruent to side $\overline{XY}$ which has U as one endpoint.

2. Label the other endpoint V.

3. Construct an angle congruent to ZXY, with vertex U and side $\overrightarrow{UV}$.

4. This angle has two sides. $\overrightarrow{UV}$ is one side.
   On the other side of the angle, lay off a segment congruent to side $\overline{XZ}$ and with U as one endpoint.

5. Label the other endpoint W.

6. Draw $\overline{VW}$.

7. Is triangle WVU congruent to triangle ZYX? _ _ _ _ _
   Triangle UVW is a <u>reproduction</u> of triangle XYZ.

Problem: *Reproduce a given triangle.*

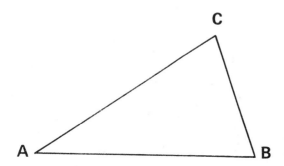

Solution:

1. Draw a line.

2. On the line, construct a segment congruent to side $\overline{AB}$. Label its endpoints D and E.

3. Draw an arc with center D and radius congruent to side $\overline{AC}$.

4. Draw an arc with center E and radius congruent to side $\overline{BC}$. Make these two arcs intersect.

5. Label their point of intersection F.

6. Draw triangle DEF.

7. Is triangle DEF congruent to triangle ABC? _ _ _ _ _

Review

1. Construct a line perpendicular to the given line through the given point on the line.

2. Bisect the given segments.

1. Bisect the given angles.

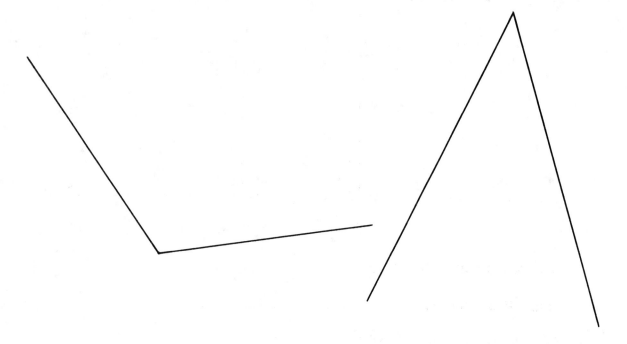

2. Divide the given segment into four congruent parts.

1. The given triangles are congruent.

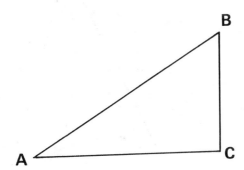 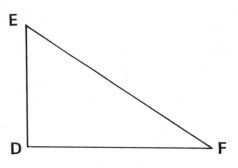

Side $\overline{AC}$ is congruent to side _ _ _ _ _ .

Side $\overline{AB}$ is congruent to side _ _ _ _ _ .

Side $\overline{BC}$ is congruent to side _ _ _ _ _ .

Angle BAC is congruent to angle _ _ _ _ _ .

Angle ACB is congruent to angle _ _ _ _ _ .

Angle ABC is congruent to angle _ _ _ _ _ .

2. Construct an angle with vertex P congruent to the given angle.

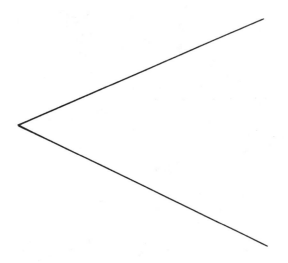

P.

Comparing the Sides and Angles of Polygons

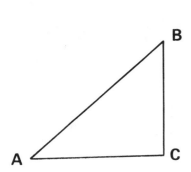 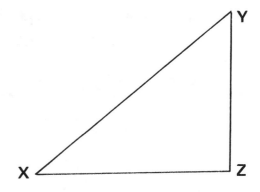

1. Check: Is angle BAC congruent to angle YXZ? _ _ _ _ _

   Check: Is angle ABC congruent to angle XYZ? _ _ _ _ _

   Check: Is angle ACB congruent to angle XZY? _ _ _ _ _

2. Check: Is side $\overline{AC}$ congruent to side $\overline{XZ}$? _ _ _ _ _

   Check: Is side $\overline{AB}$ congruent to side $\overline{XY}$? _ _ _ _ _

   Check: Is side $\overline{BC}$ congruent to side $\overline{YZ}$? _ _ _ _ _

3. Are the triangles congruent? _ _ _ _ _

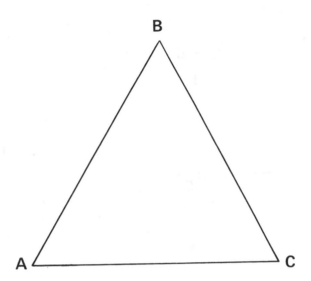

1. Construct a line perpendicular to $\overleftrightarrow{AC}$ through B.

2. Label as D the point of intersection of the perpendicular and $\overleftrightarrow{AC}$.

3. Is triangle ABD congruent to triangle CDB? _ _ _ _ _

   How do you know? _

   _

4. Compare the sides and angles of the triangles below.

   Are the triangles congruent? _ _ _ _ _

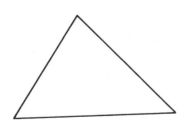
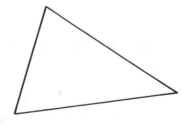

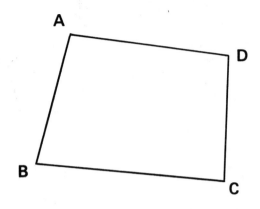

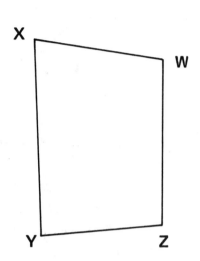

1. Check: Is angle ABC congruent to angle WXY? _ _ _ _ _ _

   Check: Is angle BCD congruent to angle XYZ? _ _ _ _ _ _

   Check: Is angle CDA congruent to angle YZW? _ _ _ _ _ _

   Check: Is angle BAD congruent to angle XWZ? _ _ _ _ _ _

2. Check: Is side $\overline{BC}$ congruent to side $\overline{XY}$? _ _ _ _ _ _

   Check: Is side $\overline{AB}$ congruent to side $\overline{WX}$? _ _ _ _ _ _

   Check: Is side $\overline{AD}$ congruent to side $\overline{WZ}$? _ _ _ _ _ _

   Check: Is side $\overline{CD}$ congruent to side $\overline{YZ}$? _ _ _ _ _ _

3. Are the quadrilaterals congruent? _ _ _ _ _ _

The quadrilaterals are congruent.

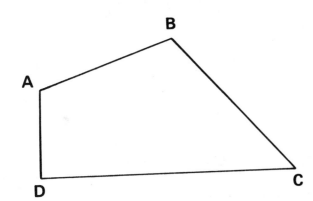

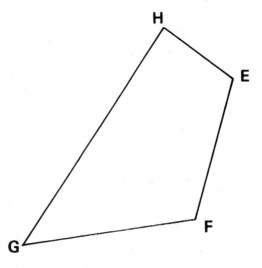

1. Side $\overline{DC}$ is congruent to side _ _ _ _ _.

   Side $\overline{AD}$ is congruent to side _ _ _ _ _.

   Side $\overline{AB}$ is congruent to side _ _ _ _ _.

   Side $\overline{BC}$ is congruent to side _ _ _ _ _.

2. Angle ADC is congruent to angle _ _ _ _ _.

   Angle DAB is congruent to angle _ _ _ _ _.

   Angle BCD is congruent to angle _ _ _ _ _.

   Angle ABC is congruent to angle _ _ _ _ _.

3. Compare the sides and angles of the quadrilaterals below.

   Are the quadrilaterals congruent? _ _ _ _ _.

Reproducing a Polygon

Problem: *Reproduce a given quadrilateral.*

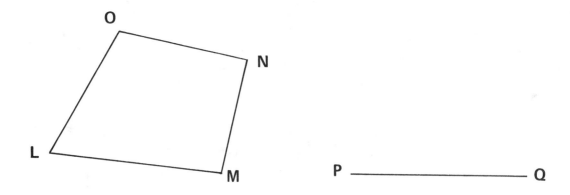

Solution:

1. Draw a line segment $\overline{PQ}$ congruent to $\overline{LM}$.

   Is the segment $\overline{PQ}$ on the page congruent to $\overline{LM}$? _ _ _ _ _

2. Reproduce the angle OLM at P with $\overrightarrow{PQ}$ as one side.

3. Construct a segment (with endpoint P) congruent to $\overline{LO}$ on the other side of the angle at P.

4. Label the new endpoint S.

5. Reproduce the angle LMN at Q with $\overrightarrow{QP}$ as one side.
   Make the other side lie above $\overrightarrow{QP}$.

6. Construct a segment (with endpoint Q) congruent to $\overline{MN}$ on the other side of the angle at Q.

7. Label the new endpoint R.

8. Draw $\overline{RS}$.

9. Is quadrilateral PQRS congruent to quadrilateral LMNO? _ _ _ _ _

Problem: *Reproduce a given quadrilateral.*

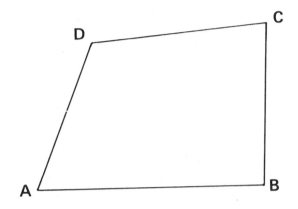

W .

Solution:

1. Draw a line segment, with endpoint W, congruent to $\overline{AB}$.
   Label the other endpoint X.

2. Reproduce the angle DAB at W with $\overrightarrow{WX}$ as one side.

3. Reproduce the angle ABC at X with $\overrightarrow{XW}$ as one side.

4. Construct a segment congruent to $\overline{AD}$ (with endpoint W) on the other
   side of the angle at W.

5. Label the new endpoint Z.

6. Construct a segment congruent to $\overline{BC}$ (with endpoint X) on the other
   side of the angle at X.

7. Label the new endpoint Y.

8. Draw $\overline{ZY}$.

9. Check: Is segment $\overline{ZY}$ congruent to segment $\overline{DC}$? _ _ _ _ _

   Check: Is angle WZY congruent to angle ADC? _ _ _ _ _

   Check: Is angle ZYX congruent to angle DCB? _ _ _ _ _

10. Is quadrilateral WXYZ congruent to quadrilateral ABCD? _ _ _ _ _

Problem: *Reproduce pentagon ABCDE.*

(Make the construction on the next page.)

Solution:

1. Draw a line segment, with endpoint V, congruent to side $\overline{AE}$.
   Label the other endpoint Z.

2. Reproduce the angle EAB at V with $\overrightarrow{VZ}$ as one side.

3. Construct a segment (with endpoint V) congruent to side $\overline{AB}$ on the other side of the angle.
   Label the new endpoint W.

4. Reproduce the angle ABC at W with $\overrightarrow{WV}$ as one side.

5. Construct a segment (with endpoint W) congruent to side $\overline{BC}$ on the other side of the angle at W.
   Label the new endpoint X.

6. Reproduce angle BCD at X with $\overrightarrow{XW}$ as one side.

7. Construct a segment (with endpoint X) congruent to side $\overline{CD}$ on the other side of the angle at X.
   Label the new endpoint Y.

8. Draw $\overline{YZ}$.

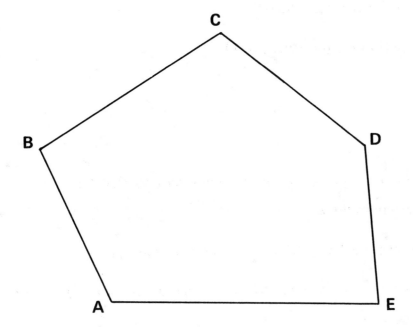

C

B

D

A

E

V ·

1. Reproduce the given triangle.

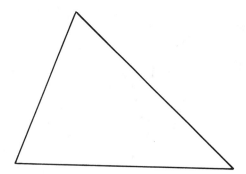

2. Reproduce the given quadrilateral.

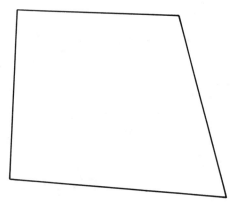

Review

1. Reproduce the given angle at X.

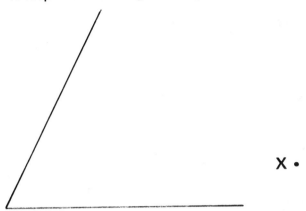

2. Construct the perpendicular to the given line through the given point.

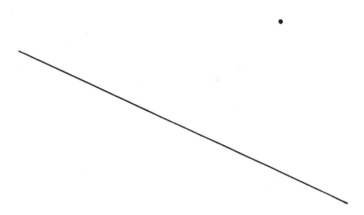

3. Bisect the angle.

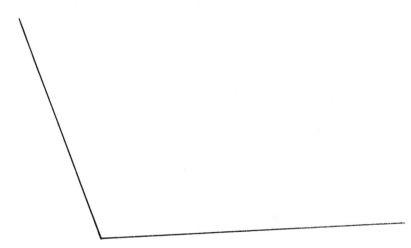

1. Reproduce the given triangle.

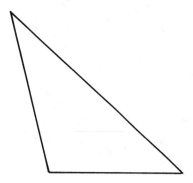

2. Which of the two angles is larger? _ _ _ _ _

 (Do not guess.)

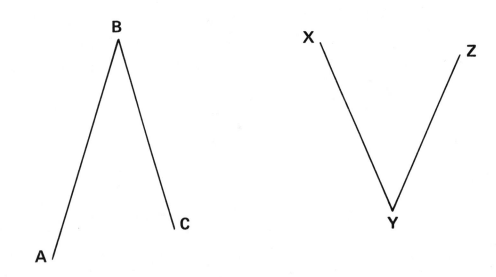

3. Construct an equilateral triangle with the given base.

1. Construct an angle, with vertex V and $\overrightarrow{VW}$ as one side, which is congruent to the given angle.

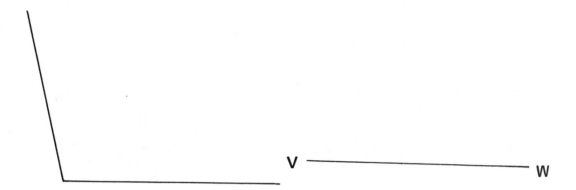

2. Triple the given segment.

_____

3. Construct a square with segment $\overline{AB}$ as one side.

A_____B

Doubling and Tripling Angles

Problem: *Construct an angle which is <u>double</u> a given angle.*

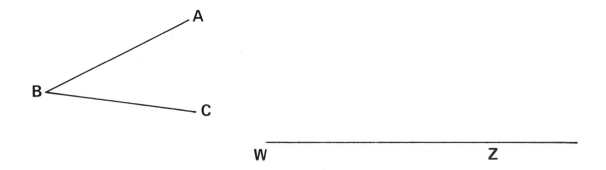

Solution:

1. Draw an arc with center B which intersects the sides of the given angle. Label the points of intersection as D and E.

2. Use the same radius to draw a large arc with center W. Label as U the point where the arc intersects ray $\overrightarrow{WZ}$.

3. Draw an arc with center U and radius congruent to segment $\overline{DE}$. Label as V the point where it intersects the arc.

4. Draw an arc with center V and radius congruent to segment $\overline{DE}$. Make it intersect the arc through U and V above V.

5. Label the point of intersection Y.

6. Draw angle YWZ.

   Is angle YWZ double angle ABC? _ _ _ _ _ _

1. Double angle DEF.

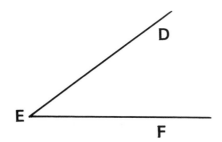

2. Triple angle KLM.

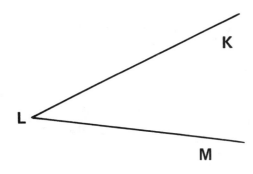

Constructing a Triangle with Sides Double those of a Given Triangle

Problem: *Construct a triangle with sides double those of the given triangle.*

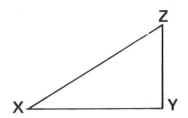

R •————————————————————————————————

Solution:

1. Draw a line segment with endpoint R which is double side $\overline{XY}$. Label its other endpoint S.

2. Double side $\overline{XZ}$.

3. Draw an arc with center R and radius congruent to the double of side $\overline{XZ}$.

4. Double side $\overline{YZ}$.

5. Draw an arc with center S and radius congruent to the double of side $\overline{YZ}$.
   Make the arcs intersect.

6. Label their intersection T.

7. Draw triangle RST.

8. Are the triangles congruent? _ _ _ _ _

   Why? _

Problem: *Construct a triangle with sides double those of the given triangle.*

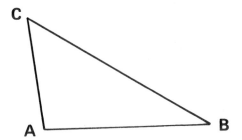

R •

Solution:

1. Draw a line segment with endpoint R which is double side $\overline{AB}$.

2. Label its other endpoint S.

3. Double side $\overline{AC}$.

4. Draw an arc with center R and radius which is congruent to double side $\overline{AC}$.

5. Double side $\overline{BC}$.

6. Draw an arc with center S and radius which is congruent to double side $\overline{BC}$.

   Make the arcs intersect.

   Label their point of intersection T.

7. Draw triangle RST.

8. Check: Is angle CAB congruent to angle TRS? _ _ _ _ _ _

   Check: Is angle ABC congruent to angle RST? _ _ _ _ _ _

   Check: Is angle ACB congruent to angle RTS? _ _ _ _ _ _

1. Construct a triangle with sides double those of the given triangle.

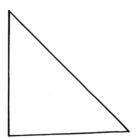

2. Construct a triangle with sides triple those of the given triangle.

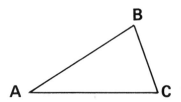

3. Label as D the point corresponding to A.

   Label as E the point corresponding to B.

   Label as F the point corresponding to C.

4. Check: Is angle EDF congruent to angle BAC? _ _ _ _ _

   Check: Is angle DEF congruent to angle ABC? _ _ _ _ _

   Check: Is angle EFD congruent to angle BCA? _ _ _ _ _

Review

1. Bisect the given angle.

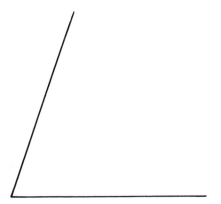

2. Bisect the segment.

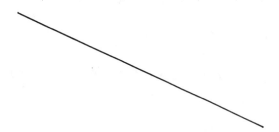

3. On the given line, construct a segment congruent to one fourth of segment $\overline{AB}$.

A ———————————— B

36

Constructing a Triangle with Sides Half those of a Given Triangle

Problem: *Construct a triangle with sides half those of the given triangle.*

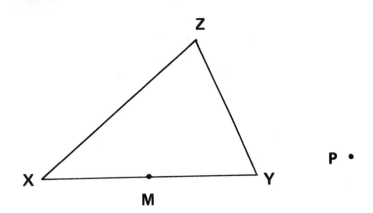

Solution:

1. Check: Is M the midpoint of side $\overline{XY}$? _ _ _ _ _

2. Draw a line segment with endpoint P which is congruent to half of side $\overline{XY}$.
   Label its other endpoint Q.

3. Bisect side $\overline{XZ}$.

4. Draw an arc with center P and radius congruent to half of side $\overline{XZ}$.

5. Bisect side $\overline{YZ}$.

6. Draw an arc with center Q and radius congruent to half of side $\overline{YZ}$.
   Make the arcs intersect.
   Label their point of intersection R.

7. Draw triangle PQR.

Problem: *Construct a triangle with sides half those of the given triangle.*

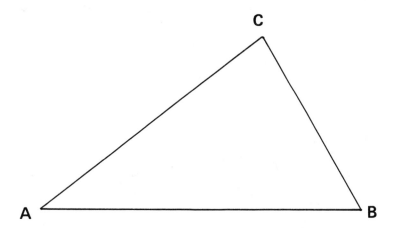

Solution: R •

1. Bisect side $\overline{AB}$.

2. Draw a line segment with endpoint R which is congruent to half of
   side $\overline{AB}$.

   Label its other endpoint S.

3. Bisect side $\overline{AC}$.

4. Draw an arc with center R and radius congruent to half of side $\overline{AC}$.

5. Bisect side $\overline{BC}$.

6. Draw an arc with center S and radius congruent to half of side $\overline{BC}$.

   Make the arcs intersect.

   Label their point of intersection T.

7. Draw triangle RST.

1. Construct a triangle with sides half those of triangle XYZ.

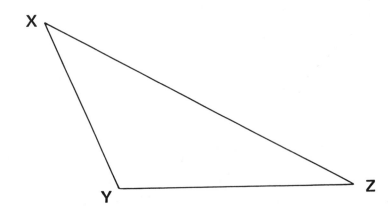

2. Label as A the point corresponding to X.

   Label as B the point corresponding to Y.

   Label as C the point corresponding to Z.

3. Check: Is angle BAC congruent to angle YXZ? _ _ _ _ _

   Check: Is angle ABC congruent to angle XYZ? _ _ _ _ _

   Check: Is angle BCA congruent to angle XZY? _ _ _ _ _

4. Are the triangles congruent? _ _ _ _ _

   Why? _

   _

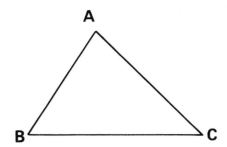

M .

1. Draw a line segment with endpoint M which is double side $\overline{BC}$.

2. Label its other endpoint N.

3. Construct an angle with vertex M and $\overrightarrow{MN}$ as one side which is congruent to angle ABC.

4. On the other side of the angle, construct a segment which is double side $\overline{AB}$.

   Label its other endpoint P.

5. Draw triangle MNP.

6. Is side $\overline{NP}$ double side $\overline{AC}$? _ _ _ _ _

Problem: *Construct a triangle with sides congruent to three quarters those of the given triangle.*

(Make the construction on the next page.)

Solution:

1. Find the midpoint of side $\overline{DE}$ and label it G.

   Find the midpoint of segment $\overline{GE}$ and label it H.

2. Draw a line segment with endpoint X which is congruent to segment $\overline{DH}$.

   Label its other endpoint Y.

   Is segment $\overline{XY}$ congruent to three quarters of side $\overline{DE}$? _ _ _ _ _

3. Find the midpoint of side $\overline{DF}$ and label it K.

   Find the midpoint of segment $\overline{KF}$ and label it L.

4. Draw an arc with center X and radius congruent to segment $\overline{DL}$.

5. Find the midpoint of side $\overline{EF}$ and lable it M.

   Find the midpoint of segment $\overline{MF}$ and label it N.

6. Draw an arc with center Y and radius congruent to $\overline{EN}$.

   Make the arcs intersect.

   Label their point of intersection Z.

7. Draw triangle XYZ.

8. Is side $\overline{XZ}$ congruent to three quarters of side $\overline{DF}$? _ _ _ _ _

   Is side $\overline{YZ}$ congruent to three quarters of side $\overline{EF}$? _ _ _ _ _

   How do you know? _

   _

   _

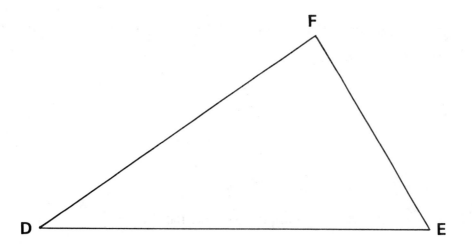

X .

Review

1. Construct a square with segment $\overline{MN}$ as one side.

M   N

2. Triple the given segment.

_____

3. Which of the two angles is larger? _ _ _ _ _

   (Do not guess.)

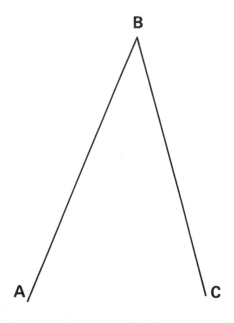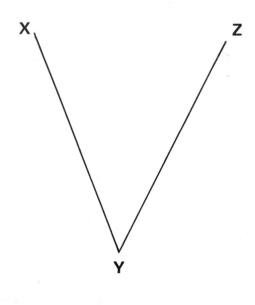

1. Construct an angle, with vertex V, which is congruent to angle XYZ.

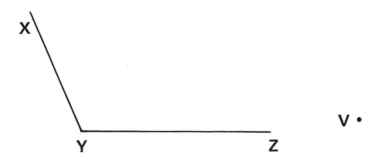

2. Reproduce the given triangle.

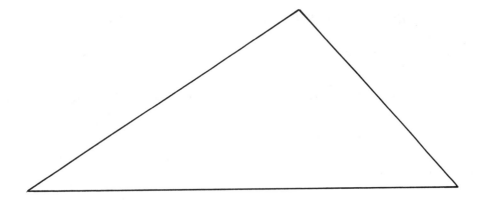

Constructing a Triangle Given the Sides

Problem: *Construct a triangle with one side congruent to segment $\overline{PQ}$, one side congruent to segment $\overline{RS}$, and one side congruent to segment $\overline{TU}$.*

P ——————————— Q   R ——————————— S   T ——————————— U

A •

Solution:

1. Draw a segment congruent to segment $\overline{PQ}$ with endpoint A.

2. Label the other endpoint B.

3. Draw an arc with center A and radius congruent to segment $\overline{TU}$.

4. Draw an arc with center B and radius congruent to segment $\overline{RS}$.
   Make the arcs intersect.

5. Label the point where the arcs intersect as C.

6. Draw $\overline{AC}$ and $\overline{BC}$.

7. Is side $\overline{AB}$ congruent to segment $\overline{PQ}$? _ _ _ _ _

   Is side $\overline{AC}$ congruent to segment $\overline{TU}$? _ _ _ _ _

   Is side $\overline{BC}$ congruent to segment $\overline{RS}$? _ _ _ _ _

Problem: *Construct a triangle ABC with side $\overline{AB}$ congruent to segment $\overline{PQ}$, side $\overline{BC}$ congruent to segment $\overline{RS}$, and side $\overline{CA}$ congruent to segment $\overline{TU}$.*

P————— Q R—————————— S T—————————U

A •

Solution:

1. Draw a segment congruent to segment $\overline{PQ}$ with endpoint A.

2. Label the other endpoint B.

3. Draw an arc with center A and radius congruent to segment $\overline{TU}$.

4. Draw an arc with center B and radius congruent to segment $\overline{RS}$.
   Make these arcs intersect.

5. Label the point where the arcs intersect as C.

6. Draw $\overline{AC}$ and $\overline{BC}$.

1. Construct a triangle with one side congruent to segment $\overline{KL}$, one side congruent to segment $\overline{MN}$, and one side congruent to segment $\overline{GH}$.

M ——————————— N   K ————————— L   G ————— H

2. Try to construct a triangle with one side congruent to segment $\overline{AB}$, one side congruent to segment $\overline{CD}$, and one side congruent to segment $\overline{EF}$.

A ————————————————— B   C ————————— D   E ———— F

3. Why is this impossible? _ _ _ _ _ _ _ _ _ _ _ _ _ _ _ _ _ _ _

_ _

_ _

Can a triangle be constructed with sides congruent to the segments given below?

A   B   C   D   E   F

Let us try!

X   Y

1. Is segment $\overline{XY}$ congruent to segment $\overline{EF}$? _ _ _ _ _
   (Don't guess.)

2. Draw an arc with center X with radius congruent to segment $\overline{CD}$.

3. Draw an arc with center Y with radius congruent to segment $\overline{AB}$.

4. Will these arcs intersect? _ _ _ _ _

1. Can a triangle be constructed with its sides congruent to any three line segments? _ _ _ _ _

2. Try to construct a triangle with sides congruent to the three given segments.

1. Try to construct a triangle with sides congruent to the three segments given below.

_____

_____

_____

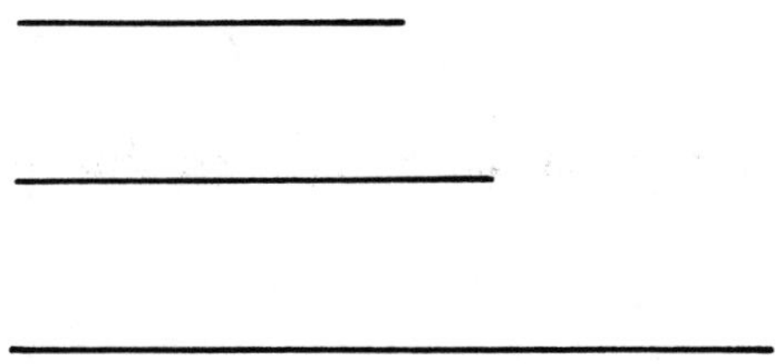

2. Try to construct a triangle with sides congruent to the three segments given below.

_____

_____

_____

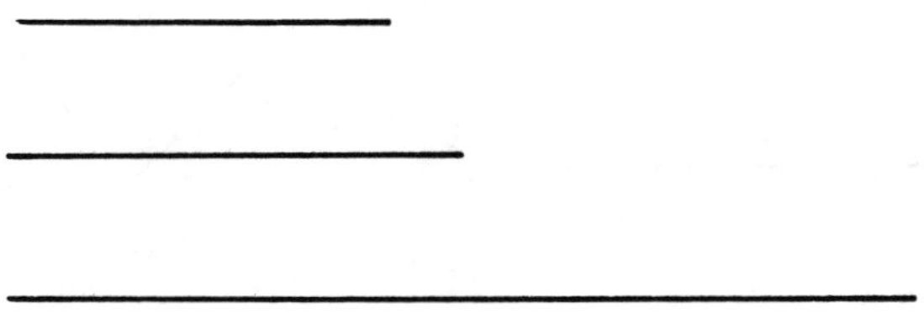

Review

1. Bisect the given line segment.

2. Are the angles congruent? _ _ _ _ _

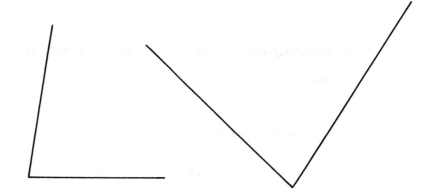

3. Reproduce the given triangle.

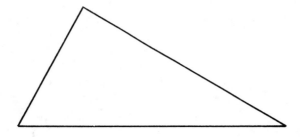

1. Construct the perpendiculars to the given lines through the given points.

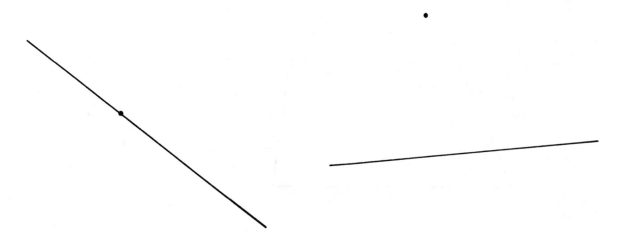

2. Bisect the given angle.

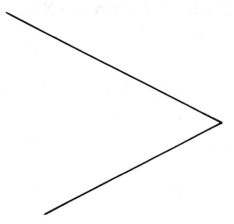

3. Construct an equilateral triangle with the given base.

1. Reproduce the given quadrilateral.

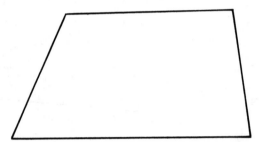

2. Construct a triangle with sides double those of XYZ.

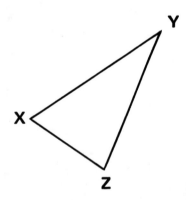

Practice Test

1. Construct an equilateral triangle with the given segment as base.

_____

2. Which segment is longer? _ _ _ _ _

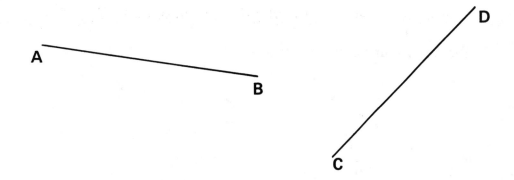

3. Construct a square with the given segment as one side.

_____

4. Construct perpendiculars to the given lines through the given points.

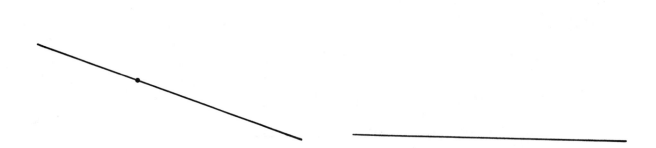

5. Bisect each angle of the triangle.

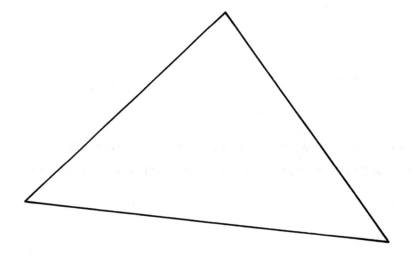

Do the bisectors meet in one point? _ _ _ _ _

6. Reproduce the given angle.

7. Reproduce the given rectangle.

8. Try to construct a triangle with sides congruent to the given segments.

Key to Geometry®

Book 1: *Lines and Segments*
Book 2: *Circles*
Book 3: *Constructions*
Book 4: *Perpendiculars*
Book 5: *Squares and Rectangles*
Book 6: *Angles*
Book 7: *Perpendiculars and Parallels,*
Chords and Tangents, Circles
Book 8: *Triangles, Parallel Lines,*
Similar Polygons
Answers and Notes for Books 1–3
Answers and Notes for Books 4–6
Answers and Notes for Book 7
Answers and Notes for Book 8

Also Available

Key to Fractions®
Key to Decimals®
Key to Percents®
Key to Algebra®
Key to Measurement®
Key to Metric Measurement®

KEY CURRICULUM PRESS
Innovators in Mathematics Education

ISBN 0-913684-76-7